HAWAIIAN
R E E F
Critters

ISLAND HERITAGE
Honolulu, Hawai'i

Contents

Corals and Their Relatives

Coral - Anemones and Zoanthids -
Hydroids - Bryozoans

Chapter Two:

Encrusting Sponges

Seaweeds

Chapter Four:

Sea Stars - Sea Urchins -
Sea Cucumbers

Acknowledgments

We would like to thank the following persons who assisted and contributed to this project:

Lu Eldredge (Pacific Science Association) and Dr. Peter Davie (Queensland Museum) who helped with the identification of some of the species.
Pauline Severns-Fiene, who provided us with last minute scientific updates on nudibranches and assisted with the identification of some of the species.
Fair Wind and Sea Paradise (Keauhou Bay), who allowed us to utilize their vessels to take photographs.
Live/Dive Pacific, whose vessels, the Kona Aggressor I & II, provided the perfect platform for ourselves, along with other contributing photographers, to learn and pursue underwater photography in Hawaii.
Smith Kenny who made an artistic contribution to the cover.
Susan Davis, who made our first three books possible and who has always shared her time and printing expertise with us as well as becoming a wonderful friend.
Keoki Stender who edited the text, and contributed his scientific knowledge, suggestions and photographs and was a pleasure to work with.

A sincere "Mahalo Nui Loa" to Fuji Film who provided us with Velvia and Provia Chrome Film for this project.

Photography:
We would like to thank the following contributing photographers:
Gui Garcia: Juvenile mushroom coral, deep water urchin, warty nudibranch, moon face side-gilled slug, pelagic sea hare, spanish dancer imperial shrimp, bull's-eye lobster, and monk seal.
Ray Mock: Boxercrab anemones, bristle worm, tiny hermit crab and red-mottled nudibranch.
Keoki Stender: Mushroom corals (Fungia and Cycloseris) and brown seaweed.
Mike and Pauline Severns: Two sea hares
John Greenamayer: Imperial shrimp on crown of thorns sea star
Glenn Fowler: Regal slipper lobster

All other photographs, including cover, by authors Astrid Witte and Casey Mahaney
© Blue Kirio, utilizing Nikon 8008S cameras in Ikelite underwater housings.

Introduction

Anyone who has had the opportunity to discover the world below, as a snorkeler or as a scuba diver, is first dazzled by the array of rainbow-hued tropical reef fish. Most of them are out in the open, swirling over the coral reef with many species even approaching snorkelers and divers with curiosity. Once the more obvious is explored, however, there is another realm of marine life to be discovered. Affectionally we refer to all life which is not classified as fish, but found on or near the coral reef, as **REEF CRITTERS**. Included within this "critter realm" are the invertebrates, animals without a backbone, along with plants, reptiles and mammals. It is amongst these critters, that some of the most spectacular and more bizarre animals are found.

Observing these critters successfully requires patience, a keen eye and at least some basic identification skills. Knowledge about the reef inhabitants' natural environment and habits will lead to a greater, much more rewarding experience. Traditionally many marine life identification books are written by scientists for scientists. Generally these books refer to the animal with only a scientific name and describe the amount of spines and markings of the particular species rather than its behavioral traits.

With divers' and snorkelers' increased awareness and interest in the reef's natural history, we have dedicated this book to fulfilling the need for an organized, easy to use identification guide for the lay person. The species listed and described in this guide are the plants, invertebrates, turtles and mammals that snorkelers and divers are most likely to encounter in Hawaii. All photographs were taken of live animals in their natural habitat. **Common Names** provide snorkelers and divers with a way of communicating in simple English. However, since common names may vary in different regions and countries, we have included the **Scientific Names** to cross reference with other marine life identification books. (If no scientific name is provided, the species has either not been scientifically described or could not be identified 100% by the photograph alone). The **Hawaiian Name** is included if available. The **Average Size** mentioned is at maturity of the species. The **Diet** consists of food the particular species normally eats. Marine life (like other wildlife) is either in its resting stage or actively pursuing prey. Being aware of its food source can assist you in finding the animal. In areas with sponge growth, for example, you can generally expect to find some nudibranchs since their diet mostly consists of sponges. However, animals often "adjust their diet" when humans interact. A Green Sea Turtle feeds naturally on algae, but will eat squid when fed by humans. (Since many critters are cryptic and little is known about their habits, we were unable to specify diet information on all species). We also provide you with information on **Where to find them** (the species) along with personal, as well as recorded scientific **Observations** on behavioral traits, when applicable.

Like Nowhere Else

The Hawaiian Islands are the most isolated islands in the world. Over 2000 miles of deep ocean separate them from the nearest land masses. Marine life depend on ocean currents to migrate to the Hawaiian archipelago while the animals are in the larval stage. At this time the plants and animals are small and buoyant and can easily drift as plankton in the ocean currents. How long they can float is determined by the species' duration of larval life. The strength and direction of the ocean currents determines how far the organisms can be carried during this time. Unfortunately, the prevailing currents don't favor Hawaii to the degree they support other parts of the Pacific. As a result, fewer species were able to reach and settle in Hawaiian waters. The existing native marine life has benefitted from a unique evolution over time. Through interbreeding, Hawaii's reefs have produced a number of species which are endemic. These fishes, shells, corals and crustaceans are found nowhere else in the world. To emphasize this unique phenomenon we have marked the endemic species with an "**E**" sign.

Reef Watching

Techniques to find the critters are as variable as the group itself. Many animals are passive and move very slowly along the bottom (sea urchins) or under ledges (nudibranchs). Others live in the dark recesses and only venture out after dark (crustaceans). Underwater Hawaii is abundant with caves and lava tubes that can support a very large and diverse number of critters. When diving or snorkeling, move slowly and carefully to avoid alerting the animals to your presence, since many critters will quickly retreat into their hideout. Many invertebrates are nocturnal, and to observe these animals while active, you will need to snorkel or dive after dark. Dusk and dawn hours also produce some excellent reef watching opportunities.

The Reef at Night

Many of the unusual and colorful critters on the reef are encountered at night. The experience of exploring the reef after dark is as rewarding for snorkelers as it is for divers. You enter yet another realm and Hawaii truly offers some of the best night diving in the world. Shrimp, lobsters, crabs, spanish dancers, beautiful cowrie shells, all rarely encountered during daytime, are frequently seen roaming the reef at night. In order to find and observe these critters it really pays to slow down. Don't try to cover as much territory as you would during the day. Often, the most interesting discoveries are made near the boat or shore entry point. There is no need to penetrate caves and lava tubes, since most of the critters will be out and about on the open reef. Most crustaceans are best observed in the outer glow of your light. Aiming the light beam directly at the animal usually causes it to retreat into a crevice.

A Note on Shell Collecting

Shell collecting has long been a popular hobby for divers, snorkelers and beach combers. Their interesting shapes and beautiful colors make for pretty treasures, but over-collecting has made many of the animals rare and disrupts the balance on the reef. Even taking dead shells affects other marine life that depend upon them for shelter and protection. If you feel you **must** collect shells, please be sure to only collect the ones with nothing living in the shell. Be careful to check them out closely because it is very common to have other marine life living in the shell. Hermit crabs, juvenile octopus and other small marine organisms take up homes in these abandoned shells. Divers and snorkelers often take shells that appear "dead" and find out later at home or in their hotel room that there was a small living creature inside. Of course, the shell starts to stink badly and ends up in the trash. A great way to collect beautiful shells for your home is to take photographs. Not only is it easier on the ecosystem, it smells better!

Symbiotic Relationships

Symbiosis is a relationship in which two organisms live in intimate association, and at least one of them benefits. Many marine animals depend on symbiotic relationships for food, transportation or other benefits necessary for survival.

Mutualism is a form of symbiotic relationships, where both organisms benefit. The Cleaner Shrimp and its host are the classic example. The host benefits, because parasites, diseased tissue and excess mucous are being removed. The shrimp, scavenging for nourishment, benefits since the host's parasites provide food for the shrimp. The shrimp represents neither prey nor predator to other organisms and its approach is readily welcomed by the host. The colorful cleaners can be observed entering the frightening mouth of a moray to deliver complete cleaning service, and although the moray could have an easy meal, it would never attempt to eat the cleaner. Featherduster Worms, which normally withdraw their fan-like extended gills with the slightest change of water motion or light, can sense the cleaner shrimp and will allow for cleaning of their delicate spirals.

The relationship between the Anemone crab and the anemone Calliactis polypus also represents a mutual symbiotic relationship, since both symbionts benefit. The crab receives protection from the anemone's stinging cells and the anemone is provided with transportation and perhaps the occasional snack. When changing to a larger shell, the hermit crab is capable of moving the anemone onto its new home.

Commensalism is formed when two symbionts enter an intimate association in which neither one will harm the other, but one may benefit from the other. The tiny Imperial Shrimp which likes to live on the Crown-of-Thorns Sea Star, benefits from the protection the venomous thorns provide. The Sea Star may not benefit from this relationship, but isn't harmed by the tiny shrimp either.

A symbiotic relationship is considered **parasitic** if one symbiont, the parasite, takes advantage of its host, such as feeding on the host's eggs or even on the host itself.

Underwater Photography

VARIOUS SYSTEMS

Before you become involved with underwater photography, you should master your diving or snorkeling skills and feel very comfortable in the water. Shallow water (no deeper than 10 feet) snap shots of reef scenes can be accomplished with waterproof disposable cameras, which are available virtually everywhere. When attempting to shoot portraits of marine animals, it is necessary to go a step further and acquire a more "serious" (i.e. expensive) system. Since color is lost at depth, you need to use an underwater strobe for your photography in order to get good results. The Nikonos and Sea & Sea cameras are two choices, which can be combined with a strobe. Both utilize a framer system for macro photography. Since the Nikonos provides superior results, we'll refer in this publication to the Nikonos framer setup, but similar images can be produced with the Sea & Sea cameras. The Nikonos can also be used with either the "close up kit" which is ideal for subjects sized at about 8 inches, or the "macro kit", which is used for subjects ranging between 1/2 inch and 2 inches. The disadvantage (of both the Nikonos and the Sea & Sea) is, that the subject needs to be placed inside the framer in order to produce a focused image and only a few animals cooperate.

For those "difficult" cases it becomes necessary to use a SLR camera (auto focus or manual focus) and place it in a specially designed underwater housing. This system has endless possibilities, but is more expensive. (See references for recommended books with more detailed information)

FISH VERSUS CRITTERS

Photographing fish generally requires a lot of patience, experience and skill. Although some fish species approach divers curiously, few are willing to "jump through the hoop" (i.e. framer) and pose with the Nikonos framer system.

With the exception of some skittish shrimps and lobsters, a majority of the reef critters are ideal beginner's subjects. Many are colorful, slow-moving creatures which don't mind having a framer placed around them. We have included **Phototips** with those critters which are the most popular photo subjects.

We have noted critters which can easily be photographed with a framer system. Please realize, that the SLR system is ALWAYS an option. With slow-moving creatures, a 55mm or 60mm macro lens tends to be the best choice, while a 100mm macro lens may be necessary for skittish, fast-moving animals, such as some of the nocturnal shrimps.

Venomous Critters

We would like to discourage you from touching any marine life, not only for your own safety, but also for the protection of the critter. Many of them are very fragile and are covered with a protective film designed to prevent infections. When handled without care or with a gloved hand, the protective layer can easily be damaged, possibly exposing the animal to infections. Some of the critters are venomous and should definitely not be touched, no matter how careful you are. We have marked the species with a "**V**" sign. There are several books available, explaining first aid and treatment of marine life injuries. Since venomous critters in Hawaii are passive, injuries are easily prevented if you refrain from touching the animal.

CHAPTER ONE

Corals and Their Relatives

Coral - Anemones and Zoanthids - Hydroids - Bryozoans

Corals and Their Relatives:

Coral - Anemones and Zoanthids - Hydroids - Bryozoans

Phylum Cnidaria ("nettle")

Coral - *Class Anthozoa*

Coral animals, the "polyps", are tiny cylindrical animals which build an external calcium carbonate skeleton around them. Each polyp has a ring of tentacles around its mouth. When touched, the polyps withdraw their tentacles and retreat into the protection of the hard skeleton. Coral growth is based on a symbiotic relationship between coral and an algae called **zooanthellae**. The growth rate is dependent on the amount of sunlight this algae receives. Abundant sunlight promotes algae growth, which in turn leads to a rapid development of the coral's calcareous skeleton. Once the reef building coral takes a firm hold, fish and other marine life can find food and shelter. Coral is the building block of the reef.

Reproduction: In some species eggs and sperm are released into the water, and fertilized eggs develop into larvae which drifts in currents as zooplankton. Other species develop larvae within their body cavities and later expel the larvae into the open water to drift. Asexual reproduction by fragmentation allows broken pieces of some species to regenerate into clones of the parent colony. Budding from a parent polyp is common in the mushroom coral. Coral colonies originate from a single polyp which undergoes division, creating hundreds of identical polyps.

COMMON NAME: Orange Cup Coral or Tubastrea Coral

SCIENTIFIC NAME: *Tubastrea coccinea*

AVERAGE SIZE: 2 inches

DIET: Zooplankton

WHERE TO FIND THEM: Grows in dark current-swept areas such as steep, undercut walls and on ceilings of caverns.

OBSERVATIONS: Most spectacular when each individual polyp opens up and extends its tentacles to feed. This is generally the case at night, or occasionally during the day, if a current is running.

COMMON NAME: Cauliflower Coral
SCIENTIFIC NAME: *Pocillopora meandrina*

AVERAGE SIZE: 12 inches
DIET: Zooplankton and nutrients provided by sunlight and algae
WHERE TO FIND THEM: Prefers shallow areas in the surge zone
OBSERVATIONS: Provides shelter for many fish, hermit, harlequin and trapezia crabs.

COMMON NAME: Antler Coral
SCIENTIFIC NAME: *Pocillopora eydouxi*

AVERAGE SIZE: 2 feet
DIET: Zooplankton and nutrients provided by sunlight and algae
WHERE TO FIND THEM: Prefers depths below the surge zone
OBSERVATIONS: Provides shelter for many fish, such as the endemic whitespot damselfish and hawkfish.

COMMON NAME: Rice Coral
SCIENTIFIC NAME: *Montipora capitata*

AVERAGE SIZE: Patches vary in size, but average about 12 inches.
DIET: Zooplankton and nutrients provided by sunlight and algae
WHERE TO FIND THEM: On the reef, often amongst other coral formations.

COMMON NAME: Finger Coral   E
SCIENTIFIC NAME: *Porites compressa*

AVERAGE SIZE: 4-6 inches (each finger)
DIET: Zooplankton and nutrients provided by sunlight and algae
WHERE TO FIND THEM: Common on the leeward sides of the Hawaiian Islands. Prefers deeper slopes and calm clear bays.
OBSERVATIONS: Finger coral provides shelter for many fish and critters.

COMMON NAME: Lobe Coral
SCIENTIFIC NAME: *Porites lobata*

AVERAGE SIZE: Variable, large mounds of 10 - 20 feet across are possible
DIET: Zooplankton and nutrients provided by sunlight and algae
WHERE TO FIND THEM: Found in shallow water. This is the most common species of coral in Hawaii.
OBSERVATIONS: Look for Christmas Tree Worms.

COMMON NAME: Plate Coral
SCIENTIFIC NAME: *Porites rus*

AVERAGE SIZE: Variable, each plate is 15 - 20 inches across.
DIET: Zooplankton and nutrients provided by sunlight and algae
WHERE TO FIND THEM: Generally below 60 feet in clear calm bays.
OBSERVATIONS: Look for the eggs of the Sergeant Major fish, a purplish layer deposited on the plate coral. The males guard their eggs against butterflyfish and wrasses, which like to feed on their eggs.

COMMON NAME: Mushroom or Razor Coral

SCIENTIFIC NAME: *Fungia scutaria*

AVERAGE SIZE: 3-5 inches
DIET: Zooplankton and nutrients provided by sunlight and algae
WHERE TO FIND THEM: Most commonly seen on rubbly areas on or near the reef.
OBSERVATIONS: Immature Mushroom Corals are attached to the bottom with a stem until they are about 1 inch in size. At this time they resemble mushrooms. As they mature they break off their stems and continue life as an unattached coral. Several new buds will grow from the base of old dead corals. This species is oblong and has wavy ribs of varying length.

COMMON NAME: Razor Coral

SCIENTIFIC NAME: *Cycloseris vaughani*

AVERAGE SIZE: 1-3 inches
DIET: Zooplankton and nutrients provided by sunlight and algae
WHERE TO FIND THEM: Found on rubbly slopes and gravel patches in water deeper than 30 feet. Uncommon species.
OBSERVATIONS: Similar to the Mushroom Coral in biology, it is always circular in outline and has six slightly thicker ribs arranged symmetrically, with smaller straight ribs in between.

COMMON NAME: Black Coral

SCIENTIFIC NAME: *Antipathes dichotoma*

AVERAGE SIZE: 6 feet
DIET: Zooplankton
WHERE TO FIND THEM: Current swept ledges deeper than 100 feet.
OBSERVATIONS: Black Coral is rare in Hawaii within the depth limit of recreational divers. Although it may grow as shallow as 30 feet, it is rarely seen in less than 100 feet, due to the fact that the shallower trees have been harvested for jewelry. Living Black Coral polyps are actually reddish-brown in color. Once removed from the water and dead, the coral skeleton looks black.

COMMON NAME: Wire Coral

SCIENTIFIC NAME: *Cirrhipathes anguina*

AVERAGE SIZE: 4 feet

DIET: Zooplankton

WHERE TO FIND THEM: Prefers current swept areas at deeper depths. Often found on deep slopes, generally distant from reef building corals such as Lobe or Finger coral.

OBSERVATIONS: Look for small Wire Coral gobies or Wire Coral shrimp (Pontonides unciger) living on the coral. Live Wire Corals appear fluorescent green until illuminated with a light, when it appears brownish-orange. The dead skeleton is dark brown and of no commercial value as jewelry.

COMMON NAME: Leather Octocoral

SCIENTIFIC NAME: *Sinularia abrupta*

AVERAGE SIZE: Variable, can cover large areas.

DIET: Zooplankton and nutrients provided by sunlight and algae

WHERE TO FIND THEM: In few locations, generally below 20 feet. Exposed to surge. Uncommon in Hawaii.

OBSERVATIONS: Looks like a hard coral but is soft to the touch and has a leathery feel to it.

COMMON NAME: Snowflake Octocoral   **E**

SCIENTIFIC NAME: *Anthelia edmondsoni*

AVERAGE SIZE: 1/4 inch

DIET: Zooplankton and nutrients provided by sunlight and algae

WHERE TO FIND THEM: Tidepools and shallow protected bays, down to about 60 feet. This species is endemic to Hawaii.

OBSERVATIONS: Grows in colonies and covers rocks like a soft lavender blanket.

Anemones and Zoanthids - *Class Anthozoa*

Sea Anemones are simple animals with a cylindrical body, which live attached to a hard surface or anchored in sand. The mouth is located in the center of the crown of tentacles which face into the water column. The tentacles have stinging cells which can irritate human skin when touched. Zoanthids are much smaller than anemones, have shorter tentacles and tend to form colonies.

Reproduction: See Coral.

COMMON NAME: Tube-Dwelling Anemone

SCIENTIFIC NAME: *Arachnanthus sp.*

AVERAGE SIZE: 5 inches
DIET: Zooplankton
WHERE TO FIND THEM: Rubble and sand patches. Often near or underneath overhangs. It is easiest to discover these animals during night dives, when their tentacles are extended to feed.

OBSERVATIONS: These nocturnal anemones live inside parchment-like tubes. The tentacles remain hidden inside the tube during the day. Tube anemones can be distinguished from other anemone-like animals by their tentacles. Long, tentacles extend from the outer edge of the oral disc, and a ring of shorter tentacles is located in the center of the disc around the mouth. It will retract into the tube if disturbed or a light is aimed directly at the animal.

COMMON NAME: Hermit Crab Anemone

SCIENTIFIC NAME: *Calliactis polypus*

AVERAGE SIZE: 2 inches
DIET: Zooplankton and scraps of food left by the hermit crab
WHERE TO FIND THEM: Attached to abandoned shells which have been adopted by hermit crabs. The best way to find them is at night when the hermit crab (Dardanus pedunculatus) roams the reef and the anemone's tentacles are extended.

COMMON NAME: Pompom or Boxer Crab Anemone

SCIENTIFIC NAME: *Bunodeopsis sp.*

AVERAGE SIZE: ¼ inch
WHERE TO FIND THEM: On the claws of the Boxer Crab (Lybia edmondsoni)

© RAY MOCK

COMMON NAME: Zoanthid

SCIENTIFIC NAME: *Palythoa sp.*

AVERAGE SIZE: ½ inch
DIET: Zooplankton
WHERE TO FIND THEM: On shallow reefs, often under ledges and overhangs and in tidepools.

Hydroids - *Class Hydrozoa ("water animal")*

Hydroids tend to live in colonies. They grow a skeleton, which resembles a feather or fern and is generally based on a structure with a stalk and branches. The polyps are attached to the branches. Hydroids have stinging cells and can cause minor to serious reaction in humans.

Reproduction: Some polyps are dedicated for reproduction and produce buds that form free-swimming medusae. At this stage they resemble tiny jellyfish and drift as zooplankton.

CORALS AND THEIR RELATIVES

COMMON NAME: Fern Hydroid

SCIENTIFIC NAME: *Pennaria disticha*

AVERAGE SIZE: 1½ inches
DIET: Zooplankton
WHERE TO FIND THEM: Out on the open reef.
OBSERVATIONS: Hydroids possess stinging cells which can cause minor to serious reactions in humans.

COMMON NAME: Feather Hydroid

AVERAGE SIZE: 1 inch
DIET: Zooplankton
WHERE TO FIND THEM: Usually underneath overhangs and in crevices.
OBSERVATIONS: Hydroids possess stinging cells which can cause minor to serious reactions in humans.

Bryozoans - *Phylum Ectoprocta ("outside anus")*
Class Gymnolaemata

Bryozoans are not directly related to corals, but due to their similar appearance are often called "Lace Coral". Like coral polyps, bryozoans build a skeleton that protect the tiny animals living inside. These animals are called zooids and differ from coral polyps by the fact that they possess a complete digestive tract.

Reproduction: See Coral.

COMMON NAME: Bryozoan or Lace Coral

SCIENTIFIC NAME: *Reteporellina denticulata*

AVERAGE SIZE: 3 inches
DIET: Phytoplankton and bacteria
WHERE TO FIND THEM: On the coral reef, in crevices or underneath overhangs.
OBSERVATIONS: Bryozoans grow in colonies. When feeding, they extend a tentacle-like feeding structure.

Sponges

Sponges
Phylum Porifera

Sponges are the simplest of all multi-celled animals and are permanently attached as adults. Their soft body is composed of interlaced fiber tissues called spongin and tiny rods, called spicules. They possess no nervous system or muscles and are very efficient filter feeders. With the help of special feeding cells, they draw water into their inner canals and filter it for microscopic organisms. Currents are helpful to facilitate water movement into their system, and due to this, Hawaii's encrusting sponges are often prominent in areas exposed to currents, the most spectacular perhaps is the crimson red sponge. Sponges produce toxin to discourage predators, which may cause irritation to humans. Avoid contact with sponges since the sharp, glassy spicules are easily embedded in the skin and very difficult to remove. In areas of high sponge growth, you may often find one of their main predators, the nudibranch.

Reproduction: Sponges have both male and female organs. They may reproduce sexually by producing eggs and sperm, or asexually by regeneration.

COMMON NAME: Encrusting Red Sponge

SCIENTIFIC NAME: *Spirastrella coccinea*

AVERAGE SIZE: Patches of up to 20 inches across

DIET: Phytoplankton and bacteria

WHERE TO FIND THEM: Under ledges, in caves and current exposed areas such as the steep outside wall of Molokini Crater.

OBSERVATIONS: Produces toxin to discourage predators.

COMMON NAME: Encrusting Orange Sponge V

AVERAGE SIZE: Small patches averaging 2 inches across
DIET: Phytoplankton and bacteria
WHERE TO FIND THEM: In shallow turbid lagoons.
OBSERVATIONS: Possibly produces toxin to discourage predators.

SPONGES

Seaweeds

Seaweeds

Phylum Chlorophyta (green algae)
Phylum Phaeophyta (brown seaweeds)

Seaweed or algae are marine plants which play an important part in the ecosystem. This group serves as food and shelter to marine animals, as well as providing them with oxygen. There are numerous species in Hawaii, enough to fill an entire book (see references). Included in our book are some of the species divers and snorkelers ask us about frequently.

Reproduction: Some species reproduce asexually by spores, with each spore capable of growing into a new seaweed. Other species reproduce sexually by gametes. A male and female gamete have to fuse to form one cell, which is then capable of growing into a new seaweed.

COMMON NAME: Green Seaweed

SCIENTIFIC NAME: *Halimeda opuntia*

HAWAIIAN NAME: Limu

AVERAGE SIZE: 4 inches

DIET: Photosynthesis

WHERE TO FIND THEM: On reef flats in calm shallow areas or amongst coral branches on the reef in deeper water.

OBSERVATION: Can range from bright green to whitish-green. Generally grows in large patches. Dead places create large deposits of "oatmeal sand" in some areas.

COMMON NAME: Seagrapes

SCIENTIFIC NAME: *Caulerpa racemosa*

HAWAIIAN NAME: Limu
AVERAGE SIZE: 2 inches (each cluster)
DIET: Photosynthesis
WHERE TO FIND THEM: In tidepools and on reef flats, often nestled amongst coral.

COMMON NAME: Brown Seaweed

SCIENTIFIC NAME: *Turbinaria ornata*

HAWAIIAN NAME: Limu
AVERAGE SIZE: 3 inches
DIET: Photosynthesis
WHERE TO FIND THEM: On shallow reef flats and tide pools.

SEAWEEDS

A Yellow Tang feeds on the Algae which grows on a Green Sea Turtle.

CHAPTER FOUR

Echinoderms

Sea Stars - Sea Urchins - Sea Cucumbers

Echinoderms

Sea Stars – Sea Urchins – Sea Cucumbers

Phylum Echinodermata ("spiny skinned")

Locomotion is accomplished with the help of a remarkable water-vascular system. Water is drawn into the system through a sieve plate causing a hydraulic system to expand and contract the animal's tubefeet with their tiny suction cups. Sea Urchins are also aided in movement by using their spines.

Sea Stars -

Class Asteroidea ("star shaped")
Class Ophiuroidea - Brittlestar

Most sea stars have five arms, but a few species have more. Their mouth is located on their underside and their anus on top. A common feeding practice occurs when the sea star everts its stomach and pours digestive enzymes over the victim. As the victim's tissues begin to disintegrate, it is being digested. Once the sea star finishes feeding, its everted stomach is drawn back into the body.

Reproduction: Sexes are generally separate and fertilization is external. Reproduction can also be accomplished by regeneration. An arm that is broken off will grow into a new complete sea star, while the sea star which dropped the arm grows a replacement.

COMMON NAME: Green Sea Star

SCIENTIFIC NAME: *Linckia diplax*

AVERAGE SIZE: 12 inches
DIET: A suspension feeder of phytoplankton, and detritus
WHERE TO FIND THEM: A fairly common sea star on the reef. Can be found in a variety of habitats ranging from coral rich areas to barren rubble and sand.

COMMON NAME: Red Spot Sea Star

AVERAGE SIZE: 1 inch
WHERE TO FIND THEM: Near the reef in rocky or rubbly areas hiding underneath rocks or in crevices.

COMMON NAME: Freckled Sea Star

SCIENTIFIC NAME: *Linckia multifora*

AVERAGE SIZE: 6 inches
DIET: A suspension feeder of phytoplankton and detritus
WHERE TO FIND THEM: Often seen on lava formations; on outer walls of caves, tunnels or arches.

COMMON NAME: Spiny Sea Star

SCIENTIFIC NAME: *Mithrodia sp.*

AVERAGE SIZE: 4 inches
DIET: A suspension feeder of phytoplankton and detritus
WHERE TO FIND THEM: Often seen on lava formations; on outer walls of caves, tunnels or arches.

COMMON NAME: Crown-of-Thorns   **V**

SCIENTIFIC NAME: *Acanthaster planci*

AVERAGE SIZE: 12 inches
DIET: Living coral
WHERE TO FIND THEM: Can be found anywhere on the coral reef. Their population varies from uncommon to population explosions, when thousands of these sea stars graze on the coral.
OBSERVATIONS: Covered with venomous sharp spines. Between the spines are small projections (papulae) which are extensions of the inner body wall and are used for respiration. Look for the tiny Imperial Shrimp (Periclimenes soror), which is often found living on the surface of the Crown-of-Thorns Sea Star. Crown-of-Thorns appear greenish in color, unless a light is used to expose the true reddish-brown color.

PHOTOTIP: Interesting shots, almost resembling a futuristic city, can be produced when taking close up images of the spines and tentacles. Photographs of the sea star are easily taken, since they move very slowly. (Remember not to touch the spines!)

COMMON NAME: Pincushion Sea Star or Shark's Pillow

SCIENTIFIC NAME: *Culcita novaeguineae*

AVERAGE SIZE: 10 inches
DIET: Living coral
WHERE TO FIND THEM: A moderately common sea star on the reef. Seen on rocky areas near coral or grazing directly on the coral.
OBSERVATIONS: Often you'll find the tiny shrimp (Periclimenes soror) living on the sea star. Pincushion sea stars feature spectacular color patterns in bright red, blue, yellow or orange.

PHOTOTIP: Because of their attractive colors, these sea stars make great photo subjects. Although it's easiest to photograph them with a close-up kit framer, stunning images can also be produced with a wide angle lens, especially when including the sun in the photograph.

COMMON NAME: Brittle Star

SCIENTIFIC NAME: *Ophiocoma pica*

AVERAGE SIZE: 5 inches
DIET: Organic matter
WHERE TO FIND THEM: On the reef and in rocky or rubbly areas hiding underneath rocks or in crevices. At night they can be observed in the open while feeding.
OBSERVATIONS: Brittle Stars are very light-sensitive and will retreat immediately when a light is directed at them. They will sacrifice an arm to save the rest of their body when threatened. This arm can be regenerated.

Sea Urchins - *Class Echinoidea* ("spiny")

The mouth is located on the underside of the body and is equipped with well-developed jaws and a set of horny teeth. The anus is located on top. The body of the urchin is basically a hollow shell made of calcium carbonate. (Called the "test"). Little space is occupied by internal organs. During the reproductive season the test is crammed with eggs or sperm.

Reproduction: Sexes are separate. Eggs and sperm are released into the water, where eggs are fertilized.

COMMON NAME: Black Spiny Urchin
SCIENTIFIC NAME: *Echinothrix diadema*

HAWAIIAN NAME: Wana (pronounced Vana)
AVERAGE SIZE: 10 inches
DIET: Seaweed
WHERE TO FIND THEM: Commonly seen in all types of habitats. During the day they stay in sheltered locations and emerge at night to feed.
OBSERVATIONS: They have needle-like spines which are extremely fine and venomous. If a diver makes contact, the spines can puncture the skin, break off in the flesh and cause a painful wound. Look for commensal urchin shrimp (Stegopontonia commensalis) within the spines of this urchin.

COMMON NAME: Deep Water Urchin
SCIENTIFIC NAME: *Astropyga radiata*

© GUI GARCIA

AVERAGE SIZE: 4 inches
WHERE TO FIND THEM: Found in deep water, but occasionally can be observed in sand patches in areas within the recreational scuba diving limits.

COMMON NAME: Black-and-White Urchin **V**

SCIENTIFIC NAME: *Echinothrix calamaris*

HAWAIIAN NAME: Wana (pronounced Vana)
AVERAGE SIZE: 11 inches
DIET: Seaweed
WHERE TO FIND THEM: Commonly seen in all types of habitats. During the day they stay in sheltered locations and emerge at night to feed.
OBSERVATIONS: They have needle-like spines which are extremely fine and venomous. If a diver makes contact, the spines can puncture the skin, break off in the flesh and cause a painful wound. Look for commensal urchin shrimp (Stegopontonia commensalis) within the spines of this urchin.

COMMON NAME: Brittle Urchin **V**

AVERAGE SIZE: 5 inches
WHERE TO FIND THEM: We have found this urchin only deep in the recess of caves. It is rare but may be seen at night in a few locations.
OBSERVATIONS: The needle-like spines are venomous and quite painful.

COMMON NAME: Slate Pencil Urchin

SCIENTIFIC NAME: *Heterocentrotus mammillatus*

AVERAGE SIZE: 10 inches
DIET: Seaweed
WHERE TO FIND THEM: Found in very shallow to deeper water, usually wedged into crevices in the coral or rocks.
OBSERVATIONS: This species has three types of spines: the long slate pencil-like spines keep predators away, the shorter ones on the underside clamp onto the reef and the flat spines, between the long triangular ones, surround and protect the body. Native Hawaiians used the large spines for writing sticks. Today the spines are often used to produce jewelry.

COMMON NAME: Rough-spined or Sputnik Urchin   E

SCIENTIFIC NAME: *Prionocidaris hawaiiensis*

AVERAGE SIZE: 10 inches
DIET: Seaweed ?
WHERE TO FIND THEM:
Found on deeper reefs in crevices near living coral. At night they emerge into the open to feed. Endemic to Hawaii.
OBSERVATIONS: In contrast to the Slate Pencil Urchin, this species has just a few large spines. The fully grown spines lose their protective skin covering and become covered with sponges, small mollusks and algae.

COMMON NAME: Rock-boring Urchin

SCIENTIFIC NAME: *Echinometra mathaei*

HAWAIIAN NAME: Ina
AVERAGE SIZE: 3 inches
DIET: Seaweed
WHERE TO FIND THEM: Most commonly found in tidepools and shallow water, where they make protective burrows in lava rocks and reef.
OBSERVATIONS: Sharp spines can cause a puncture wound but they are not venomous. May be pinkish-brown or olive green in color. The Black Rock-boring Urchin is found with this species in shallower water.

COMMON NAME: Collector or Pincushion Urchin

SCIENTIFIC NAME: *Tripneustes gratilla*

HAWAIIAN NAME: Hawa'e
AVERAGE SIZE: 5 inches
DIET: Seaweed
WHERE TO FIND THEM:
Common in shallow areas. Occurs on rubble and on living coral.
OBSERVATIONS: The spines are short and relatively weak on this species. Their tubefeet are easily visible. It appears that this species does not have a home site and keeps wandering the reefs. As the name suggests, these urchins collect debris, including seaweed, litter and we've even seen these critters clinging onto dollar notes.

Sea Cucumbers - *Class Holothuroidea* ("plant-like")

Although relatively unattractive in appearance, sea cucumbers often capture the attention of divers and snorkelers, who are wondering what the heck this sluggish, sausage-shaped creature may be. When relaxed, the bodies of sea cucumbers are generally flexible, but when disturbed they have the ability to shorten and harden their body. Most sea cucumbers feed on the organic film that coats sandy surfaces. They ingest large amounts of sand and digest the organic matter. The sand passes through the digestive tract until it is expelled through the anus, leaving a characteristic trail of sand piles behind them. Their mouth is located on one end, the anus at the other. The mouth lacks teeth and is surrounded by mop-like tentacles that aid in the feeding process.

Reproduction: Sexes are separate in most species. Eggs and sperm are released into the water. Fertilization produces free-swimming larvae.

COMMON NAME: Mottled Cucumber

SCIENTIFIC NAME: *Bohadschia paradoxa*

HAWAIIAN NAME: Loli
AVERAGE SIZE: 15 inches
DIET: Organic matter
WHERE TO FIND THEM: Various depths on sandy or rubbly bottoms.
OBSERVATIONS: When harassed or threatened these sea cucumbers are able to expel their spaghetti-like organs to discourage the predator. These organs can cause mild stinging and are extremely sticky. Once stuck to bare hands, dive equipment, etc. they are difficult to remove. The sea cucumber is able to replace these organs after a while.

COMMON NAME: Prickly Cucumber

SCIENTIFIC NAME: *Euapta godeffroyi*

AVERAGE SIZE: 15 inches
DIET: Organic matter
WHERE TO FIND THEM: Hides in crevices in shallow, sandy areas during the day and emerges at night.
OBSERVATIONS: These soft-bodied sea cucumbers are commonly mistaken for sea snakes (which do not normally exist in Hawaii) or octopus arms. They lack tubefeet and move about the reef relatively slowly, partially hidden by coral, only exposing small parts of their body. When touched, they quickly retract, and shrink to a fraction of their previous size, thus being able to hide in small reef crevices.

COMMON NAME: Spiny Cucumber

SCIENTIFIC NAME: *Stichopus chloronotus*

AVERAGE SIZE: 18 inches
DIET: Organic matter
WHERE TO FIND THEM: This firm-bodied sea cucumber is less common in Hawaii and is generally spotted in sandy or rubbly areas at deeper depths.

COMMON NAME: Black Cucumber

SCIENTIFIC NAME: *Holothuria atra*

HAWAIIAN NAME: Loli
AVERAGE SIZE: 12 inches
DIET: Organic matter
WHERE TO FIND THEM: This species is generally found in sandy or rubbly areas.
OBSERVATIONS: Its soft body is usually covered with sand.

Marine Worms

Segmented Worms - Flatworms

Marine Worms
Segmented Worms – Flatworms

Phylum Platyhelminthes ("flat worm")
Phylum Annelida ("little rings")

Segmented Worms -
Class Polychaeta ("many hairs")

The bodies of these worms are segmented by grooves outside and are divided inside by partitions. The Bristle Worm has spines along both sides for locomotion and protection. The body of the Featherduster and Christmas Tree Worms are actually hidden inside a tube that the animal builds within coral heads. What you see are the extended feathery gills which are supplied with blood vessels for breathing and covered with tiny hairs and a coat of sticky mucous to catch plankton. The plankton is then funneled into the worm's mouth.

Reproduction: Sexes are separate in most segmented worms. Eggs and sperm are simultaneously released into the surrounding water. Hatched larvae will drift as free swimming zooplankton. Many worms are able to regenerate from detached segments, either intentionally budded or accidentally injured.

COMMON NAME: Christmas Tree Worm

SCIENTIFIC NAME: *Spirobranchus giganteus*

AVERAGE SIZE: 1/2 inch (spirals)
DIET: Plankton
WHERE TO FIND THEM: Found at various depths on the reef, generally associated with Lobe Coral. A very common species which can be observed on practically any reef dive.
OBSERVATIONS: The Christmas Tree-like gill fans are sensitive to light and water motion and can be withdrawn quickly when alarmed.

PHOTOTIP: When approaching the worm, be careful not to cast a shadow over the gill fans. If the animal does withdraw it generally reappears within a minute or so. The gill fans of this tubeworm can produce spectacular photographs. Be careful not to shoot down onto the animal. Place the framer into a position to shoot horizontally for more dramatic images.

COMMON NAME: Featherduster Worm

SCIENTIFIC NAME: *Sabellastarte sanctijosephi*

HAWAIIAN NAME: Kio-po'apo'ai
AVERAGE SIZE: 4 inches (spirals)
DIET: Plankton
WHERE TO FIND THEM: Featherduster Worms are less common than Christmas Tree worms. Can be seen in tide pools and in shallow areas of the reef.
OBSERVATIONS: The gill fans are sensitive to light and water motion and can be withdrawn quickly when alarmed.
PHOTOTIP: see Christmas Tree Worm.

COMMON NAME: Spaghetti Worm

SCIENTIFIC NAME: *Loimia medusa*

HAWAIIAN NAME: Kauna'oa
AVERAGE SIZE: 2 feet (tentacles)
DIET: Organic matter
WHERE TO FIND THEM: Look in rubbly, generally less attractive-looking areas at various depths.
OBSERVATIONS: The body of this tubeworm is generally buried deep in rubble with the tentacles extended over rocks or coral. It uses its long white tentacles to gather food. Any disturbance will cause the worm to retreat its tentacles quickly.

COMMON NAME: Fire or Bristle Worm

SCIENTIFIC NAME: *Pherecardia striata*

AVERAGE SIZE: 2 - 6 inches
DIET: Small animals and organic matter
WHERE TO FIND THEM: Bristle worms are often found underneath rocks in shallow water, but sometimes can be seen free-swimming by the thousands in plankton-rich areas.
OBSERVATIONS: Their bristles cause irritation and itching when touched. When cut in half, each piece continues to live.

MARINE WORMS

Flatworms -

Class Turbellaria
Order Polycladida

Flatworms are often confused with the thicker, fleshy-nudibranchs. They have no gills and respiration occurs through the skin. Flatworms are cannibalistic, which can cause confusion during mating encounters. Locomotion is accomplished by sliding across a self-secreted mat of mucous with the help of microscopic hairs.

Reproduction: Each individual flatworm functions as a male and female sexually. Flatworms can reproduce sexually and asexually by regeneration. If severed in half the animal may develop into two complete worms.

COMMON NAME: Yellow-margin Flatworm

SCIENTIFIC NAME: *Pseudoceros ferrugineus*

AVERAGE SIZE: 3 inches

DIET: Colonial tunicates

WHERE TO FIND THEM: May be found by snorkelers and divers on or near shallow reefs, crawling on rocks or in rubbly areas, but occasionally can be observed free-swimming.

PHOTOTIP: Flatworms are one of the most popular subjects of photographers with a extension tube framer. Their color patterns are spectacular, they don't move much and good shots can be accomplished by shooting straight down onto the animal. Usually, the 1:3 framer works best.

COMMON NAME: Zebra Flatworm

SCIENTIFIC NAME: *Pseudoceros zebra*

AVERAGE SIZE: 2 inches
DIET: Colonial tunicates
WHERE TO FIND THEM: May be found by snorkelers and divers on or near shallow reefs, crawling on rocks or in rubbly areas, but occasionally can be observed free swimming.
PHOTOTIP: see Yellow-margin Flatworm

Mollusks

Shelled Snails - Sea Slugs - Octopus

Mollusks
Shelled Snails – Sea Slugs – Octopus

Phylum Mollusca ("soft body")

Mollusks are soft-bodied animals that lack a true skeleton. Many species have an external shell for protection. Mollusks are an extremely diverse group and include snails, bivalves, nudibranchs, octopi and squid.

Shelled snails -

Class Gastropoda ("stomach foot")
Subclass Prosobranchia ("forward gills")

Shelled snails possess a large muscular foot which is utilized for locomotion. Another common feature is the "radula", a tongue-like structure with specialized horny teeth used for feeding and capturing prey. In the cone shell, the radula is modified to form a single harpoon-like tooth, which is venomous. Cone shells can reach out away from their shell and harpoon their prey. All are carnivorous, with some preying on fish. Cowry shells camouflage themselves by enveloping their entire shell with their mantle, an organ that secretes calcium carbonate to build the shell and also keeps it polished. The mantle is ornamented with tentacle-like projections (papillae) which aid in camouflage and may also include sense organs.

Reproduction: The sexes are separate in most seashells and fertilization is either internal or external. Eggs are generally deposited as a ribbon on the reef. Once the larvae hatches, they begin to drift as zooplankton.

PHOTOTIP - ALL SHELLS: Shells are easy and popular subjects, since the animals are slow moving and many of them have a spectacular shell. Due to their relatively large size, many of them are best photographed with a close-up kit, or when using a SLR camera, try a 60mm macro lens.

COMMON NAME: Tiger Cowry

SCIENTIFIC NAME: *Cypraea tigris*

HAWAIIAN NAME: Leho
AVERAGE SIZE: 4 - 5 inches
DIET: Sponges and algae
WHERE TO FIND THEM: Generally lives near coral in caves and underneath ledges. At night it can often be seen out on the open reef.
OBSERVATIONS: The beautiful Tiger Cowrie is the largest, and perhaps the most spectacular cowry found in Hawaii.

COMMON NAME: Reticulated Cowry

SCIENTIFIC NAME: *Cypraea maculifera*

HAWAIIAN NAME: Leho kolea
AVERAGE SIZE: 2^1/$_2$ inches
DIET: Sponges and algae
WHERE TO FIND THEM: A common, nocturnal species, often seen out in the open at night, but can also be commonly spotted underneath rocks or underneath the ceilings of lava tubes and ledges and in holes in the coral.

COMMON NAME: Gaskoin's Cowry E

SCIENTIFIC NAME: *Cypraea gaskoini*

HAWAIIAN NAME: Leho 'ula
AVERAGE SIZE: 1 inch
DIET: Red sponge
WHERE TO FIND THEM: This species hides during the day underneath dead coral pieces, but at night time comes out into the open. Generally only seen at night.
OBSERVATIONS: Very uncommon and considered endemic to Hawaii.

COMMON NAME: Snowflake or Honey Cowry

SCIENTIFIC NAME: *Cypraea helvola.*

This Honey Cowry is deposting its egg mass.

AVERAGE SIZE: 3/4 inch
DIET: Sponges
WHERE TO FIND THEM: Under rocks in shallow rubble areas.
OBSERVATIONS: This gorgeous little cowry shows that even rubble and other less attractive areas hold interesting discoveries for the inquisitive diver or snorkeler. If you turn rocks to look for these critters, be kind enough to replace the cowry as well as the rock into its original position, as not to expose creatures living underneath the rock to predators.

COMMON NAME: Mole Cowry

SCIENTIFIC NAME: *Cypraea talpa*

AVERAGE SIZE: 3 - 5 inches
DIET: Sponges
WHERE TO FIND THEM: This uncommon species hides during the day underneath rocks in areas of coral growth, but at night emerges into the open to feed. This is the best time to find the cowrie.

COMMON NAME: Orange-Banded Cowry

SCIENTIFIC NAME: *Cypraea leviathan*

AVERAGE SIZE: 3 - 4 inches
DIET: Sponges
WHERE TO FIND THEM: This species hides during the day near coral, often in areas with Orange Cup (Tubastrea) Coral growth. At night the animal comes out in the open to feed.

MOLLUSKS

COMMON NAME: Three-toothed Conch
SCIENTIFIC NAME: *Strombus dentatus*

AVERAGE SIZE: 2 inches
DIET: Algae
WHERE TO FIND THEM: Lives in sand areas, often under rubble pieces.
OBSERVATIONS: This conch moves by jerking its muscular foot with abrupt movements into the ground, causing the shell to hop.

COMMON NAME: Pimpled Basket
SCIENTIFIC NAME: *Nassarius papillosus*

AVERAGE SIZE: 2 inches
DIET: Organic matter
WHERE TO FIND THEM: This nocturnal shell hides during the day buried in the sand. At night it is often found moving actively over rocky areas searching for food. Sometimes seen in pairs.

COMMON NAME: Helmet Shell
SCIENTIFIC NAME: *Cassis cornuta*

HAWAIIAN NAME: Pu puhi
AVERAGE SIZE: 12 inches
DIET: Sea urchins
WHERE TO FIND THEM: Helmet shells live in sandy areas, often in depths below 50 feet. Look in sand flats for protrusions emerging from the sand.
OBSERVATIONS: This large-shelled snail tends to bury itself partially into the sand. They have a large foot which can be extended outside the shell.

MOLLUSKS

49

COMMON NAME: Partridge Tun

SCIENTIFIC NAME: *Tonna perdix*

HAWAIIAN NAME: Pu'oni'oni'o

AVERAGE SIZE: 4 - 5 inches

DIET: Sea cucumbers

WHERE TO FIND THEM: This nocturnal shell buries itself in the sand during the day and can be found at night moving over the sand near coral reefs to search for its prey, the sea cucumber.

OBSERVATIONS: Locomotion is accomplished by a large foot which can be extended outside the shell.

COMMON NAME: Triton's Trumpet

SCIENTIFIC NAME: *Charonia tritonis*

HAWAIIAN NAME: Pu

AVERAGE SIZE: 15 inches

DIET: Pincushion Sea Stars, Crown-of-Thorn Sea Stars and Sea Urchins.

WHERE TO FIND THEM: Found on the reef at various depths. May be seen in shallow boulder areas or shallow and deep reefs, often underneath overhangs. Other times it can be seen on the open reef, feeding on its favorite prey, the Crown-of-Thorns Sea Star.

OBSERVATIONS:. Triton's Trumpets are one of Hawaii's largest shells.

COMMON NAME: Flea or Spotted Cone Shell V

SCIENTIFIC NAME: *Conus pulicarius*

AVERAGE SIZE: $1^1/_2$ inches
WHERE TO FIND THEM: In rubble and sandy areas, often near the reef.
OBSERVATIONS:. Cone shells have a venomous, harpoon-like "tooth", do not touch!

COMMON NAME: Leopard Cone Shell V

SCIENTIFIC NAME: *Conus leopardus.*

AVERAGE SIZE: 5 inches
DIET: Burrowing annelid worms
WHERE TO FIND THEM: Look in large sand patches at 60 feet and below. The shell is often partially covered by sand.
OBSERVATIONS:. Cone shells have a venomous, harpoon-like "tooth", do not touch!

COMMON NAME: Punchtured Miter Shell

SCIENTIFIC NAME: *Mitra sticta*

AVERAGE SIZE: $2^1/_2$ - 4 inches
DIET: Sipunculid worms
WHERE TO FIND THEM: Sandy patches near the reef in relatively shallow water.

MOLLUSKS

Sea Slugs -

Class Gastropoda ("stomach foot")
Subclass Opisthobranchia ("back gills")

This group is quite large and also variable in anatomy. Although not all species listed below technically belong to the order of Nudibranchia, they are all very closely related and for simplicity's sake many dive guides commonly refer to all Orders and Suborders of these Sea Slugs as nudibranchs. The Order Nudibranchs ("naked gills") actually consists of snail-like mollusks with external gills and no shell. Many of them are vividly colored, believed to be a warning to predators of poison accumulated in their body. Their practice of feeding on poisonous sponges makes this a plausible theory. Another common feature is the "radula", specialized horny teeth used for feeding and capturing prey. The Order Anaspidea is represented by Sea Hares, many of them possessing a small internal shell while their gills are hidden within their mantle cavity. When disturbed, they discharge a purple ink to create a "smoke screen" effect, similar to an octopus. This ink is used to confuse predators. Side-gill slugs (Order Notaspidea) have an internal shell and an external gill on one side of the animal's body.

Reproduction: Each animal functions sexually as a male and female. They leave a slime trail on the reef which provides a way for them to find each other to mate. It is best not to move them off the reef so that they have a continuous trail partners can follow. Eggs are laid in flower-like ribbons, often referred to as underwater roses (Spanish Dancer). Once hatched the larvae become free-swimming zooplankton.

PHOTOTIP - ALL NUDIBRANCHS: Nudibranchs are one of the easiest and, due to their often vivid bonbon colors, most rewarding photo subjects. Great results can be achieved with the framer system (most cases 1:2 or 1:1 works best) as well as with housed SLR cameras (100mm or 60mm macro lens).

MOLLUSKS

COMMON NAME: Spanish Dancer

SCIENTIFIC NAME: *Hexabranchus sanguineus*

AVERAGE SIZE: 12 inches
DIET: Sponges
WHERE TO FIND THEM: These large beautiful nudibranchs occur throughout the Indo Pacific. During the day, they are rarely seen on the reef. They like to hide under ledges or in lava tubes and caves, often in areas with strong water movement. During January and February Spanish Dancers occur most commonly, although they do exist throughout the year. Once spotted, divers are likely to find a Spanish Dancer at the same spot over and over again, unless the animal is disturbed and moves on. After dark is the best time to find and observe this magnificent animal. At this time they are foraging over the reef and due to their large size, are easily spotted.

OBSERVATIONS: When disturbed the Spanish Dancer swims in an undulating fashion through open water, flaring out its wide, bright red margins, resembling a Spanish dancer. Leaving the protection of the reef, however, exposes the beautiful nudibranch to predators like the moray eel. If you feel you must disturb the animal, please be kind enough to return it to the safety of the reef rather than leave it in distress in open water. The tiny shrimp Periclimenes imperator can sometimes be found attached to its body or hiding in the gills. Spanish

Eggmass of the Spanish Dancer

Dancers lay their eggs on the reef in large, red rosette-like egg masses, often mistaken by divers as an underwater flower.

PHOTOTIP: Because of their spectacular color and size, Spanish Dancers are favorites of underwater photographers. Please be considerate when photographing the animal. Rather than unnecessarily stressing the Spanish Dancer and forcing it to undulate, consider photographing it in its natural environment on the reef or lava wall.

MOLLUSKS

COMMON NAME: Warty Nudibranch

SCIENTIFIC NAME: *Asteronotus cespitosus*

© GUI GARCIA

AVERAGE SIZE: 3 inches
DIET: Sponges
WHERE TO FIND THEM: Near the shallow reef, often in rubbly or rocky zones.

COMMON NAME: Gold-Lace Nudibranch

SCIENTIFIC NAME: *Halgerda terramtuentis*

AVERAGE SIZE: 1 inch
DIET: Sponges
WHERE TO FIND THEM: This pretty little nudibranch is active during the day. Divers will often find them crawling over rocky lava walls where they feed on sponges.

COMMON NAME: Striped Nudibranch   E

SCIENTIFIC NAME: *Hypselodoris sp.*

AVERAGE SIZE: 3/4 inch
DIET: Sponges
WHERE TO FIND THEM: This uncommon species prefers shallow depths and is found only in Hawaii.

MOLLUSKS

COMMON NAME: Caramel Nudibranch

SCIENTIFIC NAME: *Glossodoris rufomarginata*

AVERAGE SIZE: 1 inch
DIET: Grey sponge
WHERE TO FIND THEM: Commonly found during the day in areas with Orange Cup (Tubastrea) Coral growth.
OBSERVATIONS: Often, these critters can be observed mating. Since each animal is male and female, the animals tend to gather for "orgies", with several animals involved in the mating process. Look for their spiral egg cases in the area.

COMMON NAME: Fried Egg Nudibranch

SCIENTIFIC NAME: *Phyllidia varicosa*

AVERAGE SIZE: 2 inches
DIET: Sponges with toxic compounds
WHERE TO FIND THEM: Perhaps the most common species in Hawaii, the Fried Egg Nudibranch is active during daylight hours and generally found on reef and lava walls.
OBSERVATIONS: These nudibranchs do not have exposed visible gills. They are known to prey on toxic sponges, and to store this toxin in their bodies. They create toxic mucous which protects them from predators.

COMMON NAME: Strawberry-Freckled Nudibranch

SCIENTIFIC NAME: *Phyllidiella pustulosa*

AVERAGE SIZE: 1¹/₂ inch
DIET: Sponges with toxic compounds
WHERE TO FIND THEM: This species prefers coral-rich areas and can be observed during daylight hours.
OBSERVATIONS: It initially appears a dull gray-green, but a light will expose the true red-pink coloration. Similar to the Fried Egg Nudibranch in shape, this species is also quite common and its gills are not visible. They also prey on toxic sponges, storing the toxin in their body, which creates a toxic mucous and protects them from predators.

COMMON NAME: Blue-Black Nudibranch

SCIENTIFIC NAME: *Phyllidiopsis sphingis*

AVERAGE SIZE: 1 inch
DIET: Sponges with toxic compounds
WHERE TO FIND THEM: This uncommon species is active during daylight hours and tends to prefer cliffs and ledges.

COMMON NAME: Fairy Nudibranch

SCIENTIFIC NAME: *Pteraeolidia ianthina*

AVERAGE SIZE: 3 - 4 inches
DIET: Hydroids and octocorals, plus nutrients produced by symbiotic algae cells.
WHERE TO FIND THEM: Occurs in a variety of environments from the shallow surge zone to more protected, deeper water. They may be found on rubble, rocks, in caves, under ledges or on coral reefs. Spotting these delicate critters takes a keen eye, since they often resemble a small colony of octocoral, hydroids or seaweed.

OBSERVATIONS: Although related to nudibranchs, they do not have exposed gills. Their frilly appendages are extensions of the digestive tract (cerata) and tipped with stinging cells to protect the animal from predators. The stinging cells can cause mild reactions in humans. To protect yourself and the delicate animal, it's best not to touch it.

MOLLUSKS

COMMON NAME: Vibrating Nudibranch E

SCIENTIFIC NAME: *Chromodoris vibrata*

AVERAGE SIZE: 1 inch
DIET: Sponges
WHERE TO FIND THEM: This relatively uncommon species is found only in Hawaii. This nudibranch tends to prefer surge-free areas and may be spotted on the open reef during the day.
OBSERVATIONS: The animals vibrate their gills, which contributed to their name.

COMMON NAME: Purple-tipped Nudibranch

SCIENTIFIC NAME: *Flabellina exoptata*

AVERAGE SIZE: 1/2 inch
DIET: Sponges
WHERE TO FIND THEM: A rare species in Hawaii, this nudibranch is more commonly found in Indonesia and the South Pacific. It may occur on the reef near or underneath ledges.

COMMON NAME: Red-mottled Nudibranch

SCIENTIFIC NAME: *Platydoris formosa*

AVERAGE SIZE: 3 inches
WHERE TO FIND THEM: A rare species in Hawaii. This one was found on a rubble patch near the reef.

MOLLUSKS

COMMON NAME: Moon Face Side-gilled Slug

SCIENTIFIC NAME: *Euselenops luniceps*

AVERAGE SIZE: 2 inch

WHERE TO FIND THEM: In deeper water (below 90 feet) on sandy or muddy bottoms or, more likely, free-swimming at night.

OBSERVATIONS: This Sea Slug is seldom observed, due to their preferred depth in an unattractive habitat. Generally, we have spotted these critters when free-swimming at night.

COMMON NAME: Orange Side-gilled Slug

SCIENTIFIC NAME: *Berthellina citrina*

AVERAGE SIZE: 2 inches

DIET: Colonial tunicates

WHERE TO FIND THEM: Divers will occasionally spot the Orange Side-gill Slugs during night dives on the reef flats or walls, often in areas associated with Orange Cup (Tubastrea) Coral growth.

OBSERVATIONS: This colorful nocturnal slug carries its exposed gills along the right side of its body, rather than on top, like most nudibranchs.

COMMON NAME: Sea Hare

SCIENTIFIC NAME: *Aplysia juliana*

AVERAGE SIZE: 2 inches

DIET: Algae

WHERE TO FIND THEM: On night dives on the open reef.

OBSERVATIONS: When disturbed, the Sea Hare discharges a purple ink to create a "smoke screen" effect, similar to an octopus. This ink is used to confuse predators.

MOLLUSKS

COMMON NAME: Sea Hare
SCIENTIFIC NAME: *Aplysia parvula*

AVERAGE SIZE: 1¹/₂ inches
DIET: Algae
WHERE TO FIND THEM: On night dives on the open reef.
OBSERVATIONS: When disturbed, the Sea Hare discharges a purple ink to create a "smoke screen" effect, similar to an octopus. This ink is used to confuse predators.

COMMON NAME: Pelagic Sea Hare
SCIENTIFIC NAME: *Stylocheilus citrinus*

AVERAGE SIZE: 2 inches
DIET: Algae
WHERE TO FIND THEM: Clinging onto algae-covered flotsam at the ocean's surface.
OBSERVATIONS: This Sea Slug is a good example that interesting critters can be found virtually anywhere. They may live on drifting abandoned fishing nets, mooring balls which may have broken loose, or any other object which is able to grow algae for this unusual Sea Hare to feed on.

COMMON NAME: Striped Swallowtail
SCIENTIFIC NAME: *Chelidonura hirundinina*

AVERAGE SIZE: ³/₄ inches
DIET: Small flatworms
WHERE TO FIND THEM: In rocky, sandy sometimes rubbly areas in shallow water of sometimes turbid lagoons.
OBSERVATIONS: This tiny Sea Slug is closest related to the Side-gill Slug and Sea Hares. It can be found in the open during daytime, but is often overlooked due to its small size.

Octopus -
Class Cephalopoda ("head and foot")
Order Octopoda

Octopus are mollusks with eight arms and no shell. They possess a highly-developed nervous system and a high degree of intelligence has been demonstrated in scientific labs. Octopus are some of the most interesting reef inhabitants to encounter. They are also capable, in contrast to many other animals, of postponing immediate gratification. Instead of consuming their prey on the spot, they often

return to their burrow with their catch. When hunting for crabs, they are known to collect several crabs before taking their prey home to eat. They use their sharp beaks to bite through hard shells and paralyze their victims. Octopus have an ink sac containing a dark fluid which is ejected when the animal feels threatened. They use this to confuse the predator. They can swim using a jet-propulsion method, but spend most of their time scuttling around on the bottom.

Reproduction: Courting and mating of the octopus occurs on the reef and can sometimes be observed by scuba divers. The male octopus possesses a modified arm, sometimes referred to as the "nuptial arm", which is used to deposit a sperm packet into the female when mating. Eggs are deposited in strands or clusters in

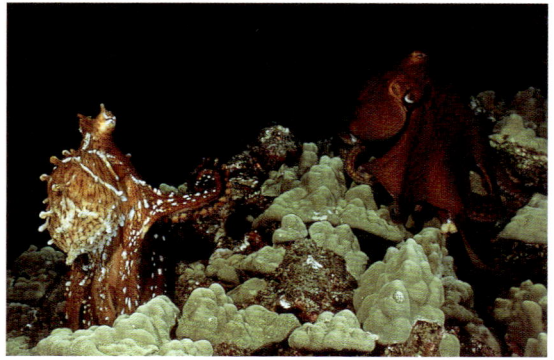

rocky crevices and are guarded by the female octopus until hatching. The incubation period takes several weeks and during this time the female octopus ceases to feed and dies shortly after her young are hatched. The young hatch looking like miniature adults. The life span of a typical reef octopus is approximately fifteen months.

COMMON NAME: Day Octopus or Tako

SCIENTIFIC NAME: *Octopus cyanea*

HAWAIIAN NAME: He'e makoko
AVERAGE SIZE: 20 inches (body and tentacles)
DIET: Crabs, shrimp, mollusks
WHERE TO FIND THEM: They can be found in a great variety of habitats. One of the easiest places to find them is in areas with coral rubble. They like to dig a burrow-like home and protect it with rocks and broken coral.
OBSERVATIONS: Day Octopus are generally curious creatures and will often sit on top of their burrow and watch over the reef for food or danger. Since they are capable of changing color and texture to blend with the environment, it generally takes an experienced observer to find them on the reef.
PHOTOTIP: Octopus are best photographed with a wide angle (35mm - 20mm) lens. Move in close for best results.

COMMON NAME: Night Octopus or Tako

SCIENTIFIC NAME: *Octopus ornatus*

HAWAIIAN NAME: He'e puloa
AVERAGE SIZE: 15 inches (body and tentacles)
DIET: Crabs, shrimp, mollusks
WHERE TO FIND THEM: In rubbly areas near the coral reef. Generally only encountered after dark.
OBSERVATIONS: This coastal species is active at night but we have seen it on rare occasions during the day on the reef. This nocturnal species is characterized by very long tentacles and white spots on the entire body.

MOLLUSKS

Gold lace nudibranch eggmass

CHAPTER SEVEN

Crustaceans

Shrimp - Mantis Shrimp - Hermit Crabs -
True Crabs - Lobsters

Crustaceans

Shrimp – Mantis Shrimp – Hermit Crabs – True Crabs – Lobsters
Phylum Arthropoda

Crustaceans belong to the Phylum Arthropoda, which is one of the largest groups in the animal kingdom. The array of members is varied and diverse and also includes insects and spiders. All of them have jointed legs as their most distinguishing feature. Many crustaceans are aggressive nocturnal hunters and hide in underwater lava tubes and caverns by day. They tend to be shy and elusive and since they come out mostly after dark, only night divers and snorkelers are able to encounter the majority of this group. Hawaii is truly one of the best places in the world to view and photograph crabs, shrimp and lobsters.

All crustaceans are protected by a rigid armor-like shell, made out of a horny substance called chitin. The external skeleton can be shed in a process called molting. This occurs when the animal goes through a growing spurt and has developed a new soft pliable shell under the hard, old one. While in this growing stage, the animal is quite vulnerable to predators and looks for a good hiding place where it can safely wait until the new shell hardens.

Some crustaceans possess claws which are used for defense or gathering food. Others possess spines on the shell for defense and long antennae for "feeling" danger.

© JOHN GREENEMAYER

Shrimp -

Class Crustacea ("covered by hard shell") *Order Decapoda* ("has ten legs")

You can often recognize shrimp by their long, hair-like antennae. Divers and snorkelers can sometimes see them sticking out from their hiding place under a ledge or coral mound during the day. The so-called "cleaning shrimp" is noted for its cleaning services. It establishes a niche on the reef that serves as a home base to clean parasites off other marine life, such as moray eels. This represents a very important service to other animals and an intricate part of the ecosystem. Cleaner shrimp are generally brightly colored and advertise their services by actively moving in an inviting manner to attract fish and other animals. Once the animal enters the cleaning area the shrimp goes to work on the scales, gills and teeth. Though it looks perilous to the shrimp, they trust their host enough to enter its mouth. It is not unusual to observe a shrimp enter the mouth of a large fish and exit out the gills. This is a classic case of symbiotic behavior and quite amazing to witness.

Reproduction: Shrimp have separate sexes, with the male impregnating the female who then carries the fertilized brood of eggs under her tail until they hatch. Once hatched, the larvae become free-swimming zooplankton.

COMMON NAME: Candycane Shrimp
SCIENTIFIC NAME: *Parhippolyte uveae*

AVERAGE SIZE: 2 inches
DIET: Organic matter
WHERE TO FIND THEM: This conspicuously-colored shrimp dwells in deep caves and lava tubes in Hawaii. It is rarely seen on the open reef, even at night.
OBSERVATIONS: This shrimp has extremely long thin legs and antennae and is often found in large groups. Directing your light towards the critters immediately causes them to wildly scurry about and to retreat even deeper into their recesses.

COMMON NAME: Scarlet Lady Cleaner Shrimp

SCIENTIFIC NAME: *Lysmata amboinesis*

HAWAIIAN NAME: 'Opae
AVERAGE SIZE: 1½ inches
DIET: Parasites and mucous of other fish
WHERE TO FIND THEM:
Although considered nocturnal, Scarlet Lady Shrimp can often be seen out in the open in daylight, when they engage in cleaning services with moray eels and other fish. Otherwise divers will spot them generally in pairs underneath ledges and in reef crevices, often waving their bright red & white antennae to advertise their services.

OBSERVATIONS: These shrimp are known to boldly approach divers to clean fingernails and even teeth, if a patient diver slowly approaches and carefully extends his hands or mouth towards the shrimp.

PHOTOTIP: Wonderful photographs can easily be taken of these shrimp while they are engaging in cleaning activities with moray or conger eels. If the eel cooperates, a framer with close-up kit could be ideal.

COMMON NAME: Barber Pole or Banded Coral Shrimp

SCIENTIFIC NAME: *Stenopus hispidus*

AVERAGE SIZE: 2 inches
DIET: Parasites and mucous coat of fish and moray eels, and organic matter
WHERE TO FIND THEM:
Although considered nocturnal, Banded Coral Shrimp can often be seen out in the open in daylight, when they engage in cleaning services with moray eels or other fish. Otherwise divers will spot them generally in pairs underneath ledges and in reef crevices.

OBSERVATIONS: This shrimp often "checks out" divers with curiosity. If a hand is extended toward them, they may boldly approach and clean the fingernails of the offered hand.

PHOTOTIP: see Scarlet Lady Cleaner Shrimp

COMMON NAME: Harlequin Shrimp

SCIENTIFIC NAME: *Hymenocera picta*

AVERAGE SIZE: 1 inch
DIET: Sea Stars
WHERE TO FIND THEM: These beautiful shrimp often live deep inside the protective branches of Cauliflower Coral. At dusk or night they can be seen out in the open searching for their favorite food, the sea star.
OBSERVATIONS: Unfortunately, Harlequin Shrimp are popular aquarium pets and have become very uncommon in their natural environment, the reef. Once spotted on the reef, Harlequin Shrimp tend to be shy but not as skittish as other shrimp. They often occur in pairs. The conspicuous coloration falsely advertises the shrimp as toxic (like nudibranchs) to discourage possible predators.

COMMON NAME: Barred Wire Coral Shrimp

SCIENTIFIC NAME: *Pontonides unciger*

AVERAGE SIZE: 1/2 inch
WHERE TO FIND THEM: This small, cryptic shrimp is found on some Wire Corals upon very careful observation.
PHOTOTIP: This tiny shrimp is easiest photographed with a 1:1 extension tube and framer. The most spectacular shots are produced after dark, when the Wire Coral polyps are open.

COMMON NAME: Transparent Wire Coral Shrimp

SCIENTIFIC NAME: *Pontonides sp.*

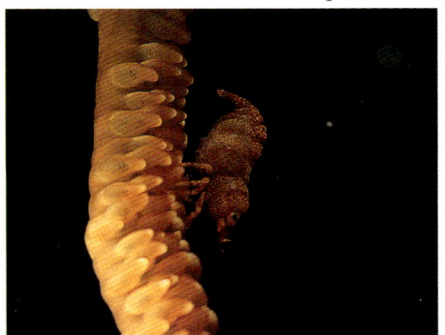

AVERAGE SIZE: 1/3 inch
WHERE TO FIND THEM: This small, cryptic shrimp is found on only a few Wire Corals upon very careful observation.
PHOTOTIP: see Barred Wire Coral Shrimp

COMMON NAME: Ghost Shrimp

SCIENTIFIC NAME: *Stenopus pyrsonotus*

AVERAGE SIZE: 4 inches
DIET: Parasites and mucous coat of other fish, organic matter
WHERE TO FIND THEM: Often seen singularly or in pairs underneath ledges during the day or at night time.
OBSERVATIONS: This species is known to clean other fish. It is also the largest shrimp found in Hawaii.

COMMON NAME: Pincushion Imperial Shrimp

SCIENTIFIC NAME: *Periclimenes soror*

AVERAGE SIZE: 1/3 inch
WHERE TO FIND THEM: This shrimp is commonly found on top or underneath Pincushion Sea Stars and Crown-of-Thorn Sea Stars, living together commensally.
OBSERVATIONS: When threatened they can move surprisingly fast, with an almost flea-like hop.

COMMON NAME: Spanish Dancer Imperial Shrimp

SCIENTIFIC NAME: *Periclimenes imperator*

AVERAGE SIZE: 1 inch
WHERE TO FIND THEM: This shrimp lives in commensalism with the Spanish Dancer nudibranch. It may be found exposed in the open on top of the nudibranch or may be nestled among the gills, or hidden in the nudibranch's mantle.

COMMON NAME: Urchin Shrimp

SCIENTIFIC NAME: *Stegopontonia commensalis*

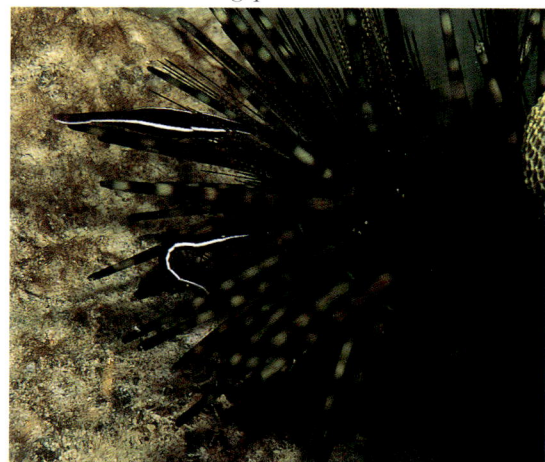

AVERAGE SIZE: 1 inch
WHERE TO FIND THEM: On the spines of Spiny Sea Urchins.
OBSERVATIONS: This shrimp lives in a commensal symbiotic relationship with the Spiny Sea Urchin. The urchin spines protect the shrimp, while the urchin receives neither harm nor benefit from the shrimp. Generally the shrimp sit in a vertical position on one of the spines, head facing down.

COMMON NAME: Marbled Shrimp

SCIENTIFIC NAME: *Saron marmoratus*

AVERAGE SIZE: 1½ inch
DIET: Organic matter
WHERE TO FIND THEM: During the day the shrimp hides in deep recesses and is rarely seen. At night they come out in the open, but always remain close to a crevice, prepared to retreat quickly.

OBSERVATIONS: This nocturnal species is very shy upon diver's approach, especially when a flashlight is used.

PHOTOTIP: This shrimp tends to be shy and retreats quickly when approached. To photograph this species with a framer is virtually impossible. Even photographers with a SLR camera should be careful not to shine a light at them until they are ready to shoot.

COMMON NAME: Exquisite Marbled Shrimp

SCIENTIFIC NAME: *Saron sp.*

AVERAGE SIZE: 1 inch
DIET: Organic matter
WHERE TO FIND THEM:
During the day the shrimp hides in deep recesses and is rarely seen. At night they come out in the open, but always remain close to a crevice, prepared to retreat quickly.
PHOTOTIP: See Marbled Shrimp

COMMON NAME: Striped Night Shrimp

SCIENTIFIC NAME: *Rhynchocinetes hiatti*

AVERAGE SIZE: 1¹/₂ inch
WHERE TO FIND THEM: Divers will usually spot the red glowing eyes of this nocturnal species, when shining the light at the shrimp during night dives. During the day the shrimp hides in deep recesses and is rarely seen.
PHOTOTIP: see Marbled Shrimp

COMMON NAME: Dotted Night Shrimp

SCIENTIFIC NAME: *Rhynchocinetes sp.*

AVERAGE SIZE: 1¹/₂ inch
WHERE TO FIND THEM: Divers will usually spot the red glowing eyes of this nocturnal species, when shining the light at the shrimp during night dives. During the day the shrimp hides in deep recesses and is rarely seen.
OBSERVATIONS: This shrimp tends to be shy and retreats quickly when approached.
PHOTOTIP: see Marbled Shrimp

Mantis Shrimp -

Class Crustacea ("covered by hard shell") *Order Stomatopoda* ("with mouth-like leg")

Mantis shrimp are very successful predators and are noted for their raptorial claws. With lightning speed they can strike out against prey and can easily slice a human finger. These shrimp are not decapods and only possess three pairs of legs.

Reproduction: Shrimp have separate sexes, with the male impregnating the female who then carries the fertilized brood of eggs under her tail until they hatch. Once hatched, the larvae become free-swimming zooplankton.

COMMON NAME: Mantis Shrimp

SCIENTIFIC NAME: *Odontodactylus scyallarus*

AVERAGE SIZE: 3 inches

DIET: Crustaceans, worms, mollusks, fish

WHERE TO FIND THEM: Feeds at night, but can be observed during the day, generally with just its head exposed in front of its burrow. This species is rare in Hawaii, but most likely to be found in areas with rubbly bottoms.

OBSERVATIONS: The Mantis Shrimp has very sharp claws which can easily slice a diver's fingers.

Hermit Crabs -

Class Crustacea ("covered by hard shell") *Order Decapoda* ("has ten legs")

Hermit crabs are actually more closely related to lobsters than crabs. These aggressive critters have adapted to live in abandoned shells. Selecting the appropriate size for their curved, flattened body is a ritual they must go through several times as they grow. Hermit crabs are known to gather at "meeting places" where they exchange shells. If two crabs choose the same shell, they will attempt to pull the shell away from the competition. Snorkelers and divers are most likely to see the front part of their body which has a protective armor with strong pincers. Most hermit crabs have one very large pincer which is used to seal off the shell's opening when threatened. The soft part of the hermit crab's body is protected by the adopted shell.

Reproduction: Sexes are separate. Fertilized eggs are carried until larvae hatch and become free-swimming zooplankton.

COMMON NAME: Orange-banded Hermit Crab
SCIENTIFIC NAME: *Trizopagurus strigatus*

AVERAGE SIZE: 3 inches
DIET: Organic matter
WHERE TO FIND THEM: They can be found in a variety of habitats but are most often spotted on or near the reef, often in small rubble areas.
OBSERVATIONS: Because of its extremely flattened body, this colorful Hermit Crab typically inhabits cone shells. These Hermit Crabs are bolder than others, and generally don't retreat entirely, making it easy to observe and photograph

COMMON NAME: Tiny Hermit Crab

SCIENTIFIC NAME: *Calcinus sp.*

AVERAGE SIZE: 1 inch

DIET: Organic matter

WHERE TO FIND THEM: This small Hermit Crab is often found within the branches of Antler Coral and Cauliflower Coral.

OBSERVATIONS: When threatened, the crab quickly retreats into its adopted shell, causing the shell to roll down the branches into the center of the coral tree for maximum shelter.

COMMON NAME: Reef Hermit Crab

SCIENTIFIC NAME: *Dardanus sp.*

AVERAGE SIZE: 3 inches

DIET: Organic matter

WHERE TO FIND THEM: During daylight hours on open reefs.

OBSERVATIONS: The abandoned unattractive shell, usually faces upward, while the hermit is hiding underneath. Divers often notice these critters, when an apparently dead shell moves quickly over the reef.

COMMON NAME: Anemone Hermit Crab
SCIENTIFIC NAME: *Dardanus pedunculatus*

AVERAGE SIZE: 4 inches
DIET: Organic matter
WHERE TO FIND THEM: Seen at night when emerging onto the open reef.
OBSERVATIONS: Anemone Crabs are nocturnal hermit crabs that can commonly be observed carrying the shell-dwelling anemone Calliactis Polypus around. This arrangement benefits both: the anemone is provided with transportation and perhaps some food on the way and the crab is protected from predators by the anemone's stinging cells.

PHOTOTIP: Anemone crabs are slow moving enough to be photographed with a close-up kit or a 1:3 macro framer. Because of the sensitive anemones attached to the shell it's best not to pick up this hermit crab. You can still get spectacular shots, if you avoid shooting down on it. Place the framer in a position which allows you to shoot horizontally at the subject.

COMMON NAME: Hairy Hermit Crab
SCIENTIFIC NAME: *Aniculus maximus*

AVERAGE SIZE: 15 inches
DIET: Organic matter and small invertebrates
WHERE TO FIND THEM: Generally found at night on the open reef or during daylight hours hiding underneath ledges or in lava tubes. This species is relatively uncommon.
OBSERVATIONS: Hairy Hermits are the largest of the Hermit Crabs and usually inhabit shells of the Triton's Trumpet or the Partridge Tun. They are surprisingly strong and have been known to pull a diver's glove off his hand.

PHOTOTIP: Due to their size and boldness these Hermit Crabs are quite easy to photograph. For large individuals a close-up wide angle shot with a 20 mm lens works well. If you use a framer be careful that the crab doesn't grab the framer and possibly pull off the lens. If you are using a close-up kit, it's best to take off the framer.

True Crabs -

Class Crustacea ("covered by hard shell")
Order Decapoda ("has ten legs")

The crab's first pair of legs has developed into two strong claws which can be used for protection, feeding and moving objects. The remaining four pairs are used for locomotion, which is performed in a quick sideways direction.

Reproduction: Sexes are separate. Fertilized eggs are carried under the abdominal plate until larvae hatch and become free-swimming zoo plankton.

COMMON NAME: 7-11 Crab
SCIENTIFIC NAME: *Carpilius maculatus*

AVERAGE SIZE: 5 inches
DIET: Mollusks and organic matter
WHERE TO FIND THEM: These large crabs are most commonly observed during night dives, but during mating season in July and August they are often spotted venturing over the open reef at daylight.
OBSERVATIONS: Named for the dark spots on its back.

COMMON NAME: Convex Pebble Crab
SCIENTIFIC NAME: *Carpilius convexus*

AVERAGE SIZE: 4 inches
DIET: Mollusks and organic matter
WHERE TO FIND THEM: During the day this crab hides under ledges and in caves and lava tubes, but during night dives this crab is commonly spotted out in the open foraging for food.

COMMON NAME: Red Reef Crab
SCIENTIFIC NAME: *Etisus splendidus*

AVERAGE SIZE: 7 inches
DIET: Small invertebrates and organic matter
WHERE TO FIND THEM: During the day this crab hides under ledges and in caves and lava tubes from the shoreline down to depths of over 100 feet. Emerges at night to feed.
OBSERVATIONS: At night, divers and even snorkelers will be able to observe and photo-graph this slow moving, heavy shelled crab.
PHOTOTIP: These bright red crabs are generally slow enough to be photographed with a close up kit. Take the framer off and just use the wand to judge distance. Avoid shining your light at the critter until you're ready to photograph.

COMMON NAME: Hawaiian Swimming Crab   E
SCIENTIFIC NAME: *Charybdis hawaiiensis*

AVERAGE SIZE: 3 inches
DIET: Small fishes, invertebrates, and organic matter.
WHERE TO FIND THEM: A very common crustacean, which lives under ledges and in crevices in the coral. At night they emerge onto the open reef.
OBSERVATIONS: Swimming crabs are very fast swimmers but can easily be observed by snorkelers and divers after dark when they venture out of their daylight hide-aways. However, if divers continue to shine a bright light at them, they will retreat into a crevice.

COMMON NAME: Sponge Crab

SCIENTIFIC NAME: *Dromia dormia*

AVERAGE SIZE: 8 - 10 inches
DIET: Mollusks, invertebrates, and organic matter
WHERE TO FIND THEM: In lava tubes, caves or underneath ledges. Look on walls and ceiling for sponge colonies.
OBSERVATIONS: These ugly-faced creatures are named after their habit of carrying sponges on their backs, holding it in place with their last pair of legs. These masters of camouflage are fairly slow moving crabs with large claws. Generally Sponge Crabs are of an earthy brown color, but when recently molted, the crab shows the new, more vulnerable pink outer layer.

COMMON NAME: Trapezia Crab

SCIENTIFIC NAME: *Trapezia rufopunctata*

AVERAGE SIZE: 3 inches
WHERE TO FIND THEM: Divers and snorkelers will spot this common crab hiding in Cauliflower and Antler Coral heads, where these little critters find shelter amongst the coral's branches.
OBSERVATIONS: Living in the coral benefits both; the crab being provided with shelter, the coral receiving protection from its predator, the Crown-of-Thorns Sea Star. If it approaches the coral, the crab will pinch the sea star's tubefeet, causing it to move on.
PHOTOTIP: Unfortunately, these attractively colored crabs are quite shy and constantly hide behind coral branches. It is difficult to get a shot with a clear view of the crab.

COMMON NAME: Decorator Crab

SCIENTIFIC NAME: *Schizophrys dama*

AVERAGE SIZE: 1 inch
WHERE TO FIND THEM: Decorator Crabs are generally active at night. The inquisitive diver who searches the reef for small organisms will occasionally spot these highly camouflaged critters on the open reef.
OBSERVATIONS: These tiny crustaceans are known to decorate themselves with pieces of coral, seaweed, algae or sponge to blend in with their environment.

COMMON NAME: Boxer or Pompom Crab   **E**

SCIENTIFIC NAME: *Lybia edmondsoni*

© RAY MOCK

AVERAGE SIZE: ³/₄ inch
WHERE TO FIND THEM: Under rocks and in reef crevices mostly in shallow areas. This species is endemic to Hawaii.
OBSERVATIONS: This rare and delicate little crab is known for its habit of carrying sea anemones (Bunodeopsis sp.) in its claws. When threatened, this crab holds the anemones up like boxing gloves for defense purposes.

COMMON NAME: Spider Rock Crab

SCIENTIFIC NAME: *Percnon planissimum*

AVERAGE SIZE: 3 inches
DIET: Algae
WHERE TO FIND THEM: Rock Crabs can be spotted crawling with surprising speed over boulders, often within the turbulent surge zone.

COMMON NAME: Striped Clown Crab

SCIENTIFIC NAME: *Xanthias sp.*

AVERAGE SIZE: 1 inch

WHERE TO FIND THEM: An undescribed species which may be seen nocturnally on the reef.

Lobsters -

Class Crustacea ("covered by hard shell") *Order Decapoda* ("has ten legs")

There are a great variety of lobsters found in Hawaiian waters. Most are claw-less and rely upon quickness and early warning from their long antennae. When they sense danger close by, they increase the movement of their sensitive antennae in order to detect the source. It is these long antennae that often alert divers to their hiding places. Slipper lobsters have a pair of thin, short antennae and a pair of flat, plate-like antennae and rely greatly on camouflage for protection.

Reproduction: Sexes are separate. Spawing occurs during summer months and fertilized eggs are held under the tail. Once hatched, the larvae become free-swimming zooplankton.

COMMON NAME: Mole Lobster

SCIENTIFIC NAME: *Palinurella wieneckii*

AVERAGE SIZE: 6 inches
DIET: Invertebrates and organic matter
WHERE TO FIND THEM: In Hawaii, divers may find this uncommon lobster hidden deep in caves and lava tubes. Generally not observed on the open reef, even at night.
OBSERVATIONS: Mole Lobsters tend to be quite skittish.

COMMON NAME: Bull's-eye Lobster

SCIENTIFIC NAME: *Hoplometopus bolthuisi*

AVERAGE SIZE: 8 inches

DIET: Various invertebrates and organic matter

WHERE TO FIND THEM: Most often found in caves, crevices and lava tubes. At night they come out in the open to look for food.

OBSERVATIONS: Due to their spectacular color they are popular subjects for macro-photographers. Unfortunately they are very uncommon.

PHOTOTIP: These relatively small reef lobsters fit into a close-up kit framer, but since they are quite skittish, it's best to take the framer off. They are easier photographed with a reflex camera. Avoid shining your light at them until you're ready to take the photo.

COMMON NAME: Hawaiian Lobster

SCIENTIFIC NAME: *Enoplometopus occidentalis*

AVERAGE SIZE: 8 inches

DIET: Various invertebrates and organic matter

WHERE TO FIND THEM: This striking lobster is most often found in caves, crevices and lava tubes. At night they come out in the open to look for food.

OBSERVATIONS: This species is related to the Maine Lobster and has real claws. Due to their spectacular color they are popular subjects for macro-photographers. Unfortunately they are relatively uncommon.

PHOTOTIP: see Bull's-eye Lobster

COMMON NAME: Hawaiian Spiny Lobster **E**

SCIENTIFIC NAME: *Panulirus marginatus*

HAWAIIAN NAME: Ula
AVERAGE SIZE: 16 inches
DIET: Various marine life and organic matter
WHERE TO FIND THEM: Spiny lobsters hide under ledges and in caves during the day and roam the reef at night. This species is endemic to Hawaii.
OBSERVATIONS: This species has no claws, but strong antennae and their body is covered with spines. When threatened, spiny lobsters use their powerful tails to propel themselves backwards.
PHOTOTIP: Because of their long antennae it's very difficult to fit the entire animal into the framer. Even for a close-up portrait it's better to take the framer off, since the antennae are sensitive to touch and will almost certainly alarm the animal to retreat. For a whole animal photo, the 20mm lens is a good choice for a wide angle close-up shot.

COMMON NAME: Tufted Spiny Lobster

SCIENTIFIC NAME: *Panulirus penicillatus*

HAWAIIAN NAME: Ula
AVERAGE SIZE: 16 inches
DIET: Algae and various animal life
WHERE TO FIND THEM: Hide under ledges and in caves during the day and roam the reef at night
OBSERVATIONS:. They have no claws, but strong antennae and their body is covered with spines. When threatened spiny lobsters use their powerful tails to propel themselves backwards.
PHOTOTIP: see Hawaiian Spiny Lobster

COMMON NAME: Long-handed Lobster
SCIENTIFIC NAME: *Justitia longimanus*

HAWAIIAN NAME: Ula
AVERAGE SIZE: 6 inches
DIET: Various invertebrates
WHERE TO FIND THEM: Found in crevices in the reef, holes in the outer walls of lava tubes and caves or inside caves, but emerges at night to feed. Lives at greater depths, generally past 60 feet.
OBSERVATIONS: This species is relatively rare in Hawaii. They have no claws, but long antennae.
PHOTOTIP: see Hawaiian Spiny Lobster

COMMON NAME: Sculptured Slipper Lobster
SCIENTIFIC NAME: *Parribacus antarcticus*

HAWAIIAN NAME: Ula papapa
AVERAGE SIZE: 8 inches
DIET: Mollusks and other invertebrates, and organic matter
WHERE TO FIND THEM: Commonly spotted at night on the open reef, or during daylight hours on or near rocks on the coral reef. Divers often find this Slipper Lobster inside caves on walls and ceilings.

PHOTOTIP: The Sculptured Slipper Lobster is easily photographed with a large framer close-up kit or with a wide angle close up method utilizing the 20mm lens.

COMMON NAME: Regal Slipper Lobster   **E**

SCIENTIFIC NAME: *Arctides regalis*

© GLENN FOWLER

HAWAIIAN NAME: Ula papapa
AVERAGE SIZE: 6 inches
DIET: Various invertebrates
WHERE TO FIND THEM: This species is more commonly found in deeper water and the outer edge of the reef, and due to this, is mostly seen by scuba divers. During day time they tend to hide underneath ledges and reef crevices but can be seen at night roaming the reef. Only found in Hawaii.

OBSERVATIONS: This species is the most spectacular of the Slipper Lobsters and due to this, is a popular photographer's subject.

COMMON NAME: Shovelnose Slipper Lobster

SCIENTIFIC NAME: *Scyllarides squammosus*

HAWAIIAN NAME: Ula papapa
AVERAGE SIZE: 10 - 11 inches
DIET: Various invertebrates
WHERE TO FIND THEM: May be spotted on the walls or ceilings of caves or lava tubes or on the open reef at night.

OBSERVATIONS: This is the largest species of Hawaii's Slipper Lobsters.

Large Critters
Aquatic Mammals - Sea Turtles

Large Critters
Aquatic Mammals - Sea Turtles
Phylum Chordata

Aquatic Mammals -

Class Mammalia

Dolphins and seals are warm blooded, air-breathing animals that make their home in the aquatic realm. These mammals had to adapt to severe physical factors, such as loss of body heat, movement in the underwater environment and staying submerged for long periods. Dolphins (and whales) are the most well-adapted mammals for life in water. Monk Seals are also excellent swimmers, who can crawl on land too.

Reproduction: Aquatic mammals give birth to live young which they suckle just as their terrestial cousins do. During courtship, Dolphins swim playfully together and often touch each other with their flippers and brush against each others body. Mating occurs in the water in a frontal position, with the animals being on their sides with respect to the surface of the water. Dolphins also give birth in the water, with the calf emerging tail first, to prevent them from drowning in case of any complications in the birthing process. The calves are fully developed.

Courting Monk Seals often muzzle nudge while lying around on the beach or they swim and play together in the water. Contrary to other seals, mating occurs in the water and it is common for the male to bite the female's back during mating. To give birth the Monk Seal mom finds a suitable place on the beach, usually in a protected bay, where the pup can slowly adjust to swimming.

Common Name: Spinner Dolphin
SCIENTIFIC NAME: *Stenella longirostris*

HAWAIIAN NAME: Nai'a
AVERAGE SIZE: 5 feet
DIET: Fish, Flying Fish, squid
WHERE TO FIND THEM: This coastal species is generally observed when playfully jumping and skillfully spinning at the surface. At night they gather in groups and swim to offshore feeding areas. During the day they return to protected near shore areas.

OBSERVATIONS: Most active at night when they school up and hunt in deep offshore waters. Spinners are quite shy and seldom approach a diver underwater.

Common Name: Whitespotted Dolphin
SCIENTIFIC NAME: *Stenella attenuata*

HAWAIIAN NAME: Nai'a
AVERAGE SIZE: 5 feet
DIET: Fish, squid
WHERE TO FIND THEM: This offshore species is commonly observed at the surface, often up to 6 miles from shore, when escorting boats and playfully surfing the wake. To hunt, the dolphins prefer the deeper offshore water, where they school up and hunt for tuna and other fish.

OBSERVATIONS: Whitespotted Dolphins are known to curiously approach snorkelers and, at times, even scuba divers, when venturing close to the shallow coastal waters.

Common Name: Hawaiian Monk Seal   E

SCIENTIFIC NAME: *Monachus schauinslandi*

© GUI GARCIA

HAWAIIAN NAME: 'Ilio-holo-i-kauaua (meaning "dog running in toughness/waves")
AVERAGE SIZE: 5 feet
DIET: Reef fish, octopus, spiny lobsters
WHERE TO FIND THEM: Most sightings occur near their breeding ground, the Northwestern Islands. Occasionally, they can be seen throughout the Hawaiian Islands. There have been several recent sightings along the coast of South Kona and Kau on the Big Island. Monk Seals like nearshore waters and are sometimes encountered on the reef. When on a boat, keep your eyes open at deserted beaches. Seals like to bask in the sun on sandy beaches.

OBSERVATIONS: Hawaiian Monk Seals are endemic to Hawaii and are an endangered species. At this time there are approximately 1200 seals left. Monk Seals are capable of diving down to 500 feet and can stay under water for up to 20 minutes.

Sea Turtles - *Class Reptilia*

Sea turtles are aquatic reptiles that are cold blooded. Due to this, they prefer the warmer zones such as Hawaii. They do not possess gills and need to return to the surface to breathe. Sea Turtles have more flexible and streamlined shells than their terrestrial cousins and their heads cannot be pulled back into the shell.

Reproduction: To lay their eggs turtles return to the beach where they were born. Males and females mate in the water near the beach. At night the female crawls high up the beach to lay up to 100 leathery golfball-sized eggs. In Hawaii the Green Sea Turtles breed and lay eggs in the Northwestern Hawaiian Islands.

Common Name: Green Sea Turtle

SCIENTIFIC NAME: *Chelonia mydas*

This turtle is being cleaned at a cleaning station.

AVERAGE SIZE: 4 feet or up to 300 pounds.
DIET: Algae
WHERE TO FIND THEM: The popular reef inhabitants are usually seen either resting on rubble bottom and finger coral reefs or swimming in midwater near the reef, usually in depths between 30 and 60 feet. They may occur as shallow as the surface or even above the surface sunning themselves on a rock, but can also be occasionally encountered at depths below 60 feet.

OBSERVATIONS: Once an endangered species, Green Sea Turtles are sighted more and more frequently in Hawaii and encounters are quite common, but remain protected by law from being disturbed or even touched. Night divers sometimes encounter turtles sleeping in a crevice on the reef. At this time, please be extra careful not to disturb the animal. Turtles can easily become disoriented at night, often leaving the protection of the reef, making them easy prey to their predator, large sharks such as the Tiger Shark.

Common Name: Hawksbill Turtle

SCIENTIFIC NAME: *Eretmochelys imbricata*

AVERAGE SIZE: 2$\frac{1}{2}$ feet or up to 200 pounds
DIET: Sponges, tunicates
WHERE TO FIND THEM: Occurs near or on the coral reef, but unfortunately, this species is endangered and rarely seen in Hawaiian waters.

Endemic Species

Venomous Marine Life

About The Authors

With a degree in education, Casey Mahaney and his partner Astrid Witte have a combined twenty years of experience in the dive and snorkel industry. After issuing over a thousand scuba certifications and introducing tens of thousands of reef watching enthusiasts to the intricacies of the reef, they are well aware of the marine life information divers and snorkelers seek. As enthusiastic underwater naturalists and photographers they have made it their goal to provide literature on marine life identification and behavior -- specializing in a style that is convenient, easy to understand, as well as entertaining.

Their photography and dozens of articles have been published in magazines worldwide, including Aqua Geographica, Skin Diver, Ocean Realm, Sportdiving, Aquanaut, Tauchen and many others. They are also the authors of *Reefwatchers Hawaii, Reefwatchers Guam & Micronesia, Hawaiian Reef Fish - The Identification Book*, and *The Essential Guide to Live-Aboard Dive Travel*. Many of their photographs are on display in art galleries throughout the Hawaiian Islands.

Their company, **Blue Kirio Travel**, offers scuba adventure tours to remote and exciting Pacific destinations, all personally escorted by Casey and Astrid. For up-to-date schedules and information, contact the authors at:

BLUE KIRIO TRAVEL
74-5602 Alapa Street # 764
Kailua-Kona, HI 96740
(800) 863 2524
e-mail: caseym@interpac.net
http://www.bluekirio.com

Index

Shelled Snails

Shrimp

Sponges

Zoanthids

References

Allen G. R. & R. Steene. *Coral Reef Field Guide.* Singapore: Tropical Reef Research, 1994.

Anon. "Hawaiian Sea Turtle." *National Marine Fisheries Service pamphlet.* Honolulu: Hawaii Division of Aquatic Resources, 1996.

Bertsch H. and S. Johnson. *Hawaiian Nudibranchs.* Honolulu: Oriental Publishing Co, 1981.

Brusca, R.C. and G. J. Brusca. *Invertebrates.* Sunderland: Sinauer Associates, Inc., 1990.

Burgess, C.M. *Cowries of the World.* Cape Town: Gordon Verhoef-Seacomber Publications, 1985.

Caldwell, D.K. and M.C. Caldwell. *The Audubon Society Field Guide to North American Fishes, Whales, and Dolphins.* New York: Alfred A. Knopf, 1983.

Church J. *Essential Guide to Nikonos Systems.* New York: Aqua Quest Publications, 1994.

Colin P.L. and C. Arneson. *Tropical Pacific Invertebrates.* Beverly Hills: Coral Reef Press, 1995.

Devaney, D. M. and L. G. Eldredge. *Reef and Shore Fauna of Hawaii, Section 1: Protozoa through Ctenophora.* B. P. Bishop Museum Special Publication 64 (1). Honolulu: Bishop Museum Press, 1977.

Devaney, D. M. and L. G. Eldredge. *Reef and Shore Fauna of Hawaii, Section 2: Platyhelminthes through Phoronida and Section 3: Sipuncula through Annelida.* B. P. Bishop Museum Special Publication 64 (2 and 3). Honolulu: Bishop Museum Press, 1977.

Fielding, A. *Hawaiian Reefs and Tidepools.* Honolulu: Oriental Publishing Co., 1985.

Fielding, A. and E. Robinson. *An Underwater Guide to Hawaii.* Honolulu: University of Hawaii Press, 1987.

Gosliner, T. M., D. W. Behrens, and G. C. Williams. *Coral Reef Animals of the Indo-Pacific.* Monterey: Sea Challengers, 1996.

Humann, P. *Reef Creature Identification Florida, Caribbean, Bahamas.* Jacksonville: New World Publications, 1992.

Kay, E. A. *Hawaiian Marine Shells. Reef and Shore Fauna of Hawaii, Section 4: Mollusca.* B. P. Bishop Museum Special Publication 64 (4). Honolulu: Bishop Museum Press, 1977.

Magruder, W.H. and J.W. Hunt. *Seaweeds of Hawaii.* Honolulu: The Oriental Publishing Co, 1979.

Russo, R. *Hawaiian Reefs.* San Leandro: Wavecrest Publications, 1994.

Witte, A. and C. Mahaney. *Reefwatchers Hawaii, Fish & Critter I.D.* Kailua-Kona: Blue Kirio Publishing, 1996.

Wilson, R. and J.Q. Wilson. *Watching Fishes.* Houston: Gulf Publishing Company, 1992.

Wu, N. *How to Photograph Underwater.* Mechanicsburg: Stackpole Books, 1994.